Heba Aboul Ezz
Yasser Khadrawy
Iman Mourad

Health Hazards of Bisphenol A

Heba Aboul Ezz
Yasser Khadrawy
Iman Mourad

Health Hazards of Bisphenol A

The Effect of Bisphenol A on Vital Organs of Adult Rat

LAP LAMBERT Academic Publishing

Impressum / Imprint
Bibliografische Information der Deutschen Nationalbibliothek: Die Deutsche Nationalbibliothek verzeichnet diese Publikation in der Deutschen Nationalbibliografie; detaillierte bibliografische Daten sind im Internet über http://dnb.d-nb.de abrufbar.
Alle in diesem Buch genannten Marken und Produktnamen unterliegen warenzeichen-, marken- oder patentrechtlichem Schutz bzw. sind Warenzeichen oder eingetragene Warenzeichen der jeweiligen Inhaber. Die Wiedergabe von Marken, Produktnamen, Gebrauchsnamen, Handelsnamen, Warenbezeichnungen u.s.w. in diesem Werk berechtigt auch ohne besondere Kennzeichnung nicht zu der Annahme, dass solche Namen im Sinne der Warenzeichen- und Markenschutzgesetzgebung als frei zu betrachten wären und daher von jedermann benutzt werden dürften.

Bibliographic information published by the Deutsche Nationalbibliothek: The Deutsche Nationalbibliothek lists this publication in the Deutsche Nationalbibliografie; detailed bibliographic data are available in the Internet at http://dnb.d-nb.de.
Any brand names and product names mentioned in this book are subject to trademark, brand or patent protection and are trademarks or registered trademarks of their respective holders. The use of brand names, product names, common names, trade names, product descriptions etc. even without a particular marking in this work is in no way to be construed to mean that such names may be regarded as unrestricted in respect of trademark and brand protection legislation and could thus be used by anyone.

Coverbild / Cover image: www.ingimage.com

Verlag / Publisher:
LAP LAMBERT Academic Publishing
ist ein Imprint der / is a trademark of
OmniScriptum GmbH & Co. KG
Heinrich-Böcking-Str. 6-8, 66121 Saarbrücken, Deutschland / Germany
Email: info@lap-publishing.com

Herstellung: siehe letzte Seite /
Printed at: see last page
ISBN: 978-3-659-74178-4

Copyright © 2015 OmniScriptum GmbH & Co. KG
Alle Rechte vorbehalten. / All rights reserved. Saarbrücken 2015

Health Hazards of Bisphenol A

The effect of bisphenol A on vital organs of adult rat

Heba S. Aboul Ezz[a], Yasser A. Khadrawy[b] and Iman M. Mourad[a]

[a]Zoology Department, Faculty of Science, Cairo University, Giza, Egypt.
[b]Medical Physiology Department, Medical Division, National Research Center, Giza, Egypt.

Bisphenol A (BPA) is an endocrine-disrupting chemical that is widely incorporated in the manufacture of polycarbonated plastics and epoxy resins and lines metal food and beverage cans. The extensive use of BPA-containing products leads to a high global population exposure which starts early during the fetal life, postnatal life and extends throughout the life of the individual. High temperature, repeated washing of polycarbonate products and exposure to either acidic or basic conditions accelerate the hydrolysis of polymers containing BPA molecules leading to an increase in the rate of leaching of BPA. The detection of low levels of BPA in over 90% of human urine samples in the US and other countries raised concerns regarding BPA's potential health effects and suggested improved study designs and methodologies to address these concerns. Many agencies raised warnings against the excessive use of such substances. Several authorities in Canada, Australia and France banned the use of BPA from baby bottles and food containers and products for very young children. To toxicologists, this book provides an overview on the health hazards of BPA focusing on the adverse effects of BPA on vital organs of adult rats. The role of oxidative stress in the toxic actions of BPA in the liver, kidney and testis and its relation to liver and kidney functions were evaluated. In addition, the cardiotoxic and neurotoxic effects of BPA were discussed.

Contents

List of abbreviations 5
1. Introduction 6
1.1. Structure of bisphenol A 6
1.2. Uses of bisphenol A 7
1.3. Bisphenol A exposure 8
1.4. Current opinion of scientific organizations 9
1.5. Mechanism of Action 11
1.6. Pharmacokinetics and metabolism 12
1.7. Health Hazards of Bisphenol A 13
1.7.1. Effects on metabolism 14
1.7.2. Effects on the reproductive system 15
1.7.3. Effects on vital organs 16
1.7.4. Effects on the cardiovascular system 17
1.7.5. Effects on the central nervous system 19
2. Current Research on the effect of bisphenol A in adult rats .. 21
2.1. Experimental Design 22
3. Discussion 23
3.1. Effect of BPA on the liver 23
3.2. Effect of BPA on the testis 26
3.3. Effect of BPA on the kidney 27
3.4. Effect of BPA on the heart 28
3.5. Effect of BPA on the body weight gain 32

3.6. Effect of BPA on the nervous system ……………………….. 33
4. Conclusion ...…………………………………………………. 39
5. Recommendation ……………………………………………. 40
6. References …………………………………………………… 42

List of abbreviations

Acetylcholinesterase (AChE)

Alanine aminotransferase (ALT)

Aspartate aminotransferase (AST)

Bisphenol A (BPA)

Blood pressure (BP)

Calcium ions (Ca^{2+})

Cardiovascular diseases (CVDs)

Central nervous system (CNS)

Endothelial nitric oxide synthase (eNOS)

Estrogen receptor α (ERα) and ERβ

Glutathione-S-transferase (GST)

Heart rate (HR)

Malondialdehyde (MDA)

National Health and Nutrition Examination Survey (NHANES)

Nitric oxide (NO)

N-methyl-D-aspartate (NMDA)

Reactive oxygen species (ROS)

World Health Organization (WHO)

γ-aminobutyric acid (GABA)

1. Introduction

1.1. Structure of bisphenol A

Bisphenol A (BPA) is a carbon-based synthetic compound with the chemical formula $(CH_3)_2C(C_6H_4OH)_2$. It belongs to the group of diphenylmethane derivatives and bisphenols, with two hydroxyphenyl groups. It is a colorless solid soluble in organic solvents, but poorly soluble in water. BPA was first synthesized in 1891, by the Russian chemist Aleksandr P. Dianin, who combined phenol with acetone in the presence of an acid catalyst to produce the chemical. In the 1950s, scientists discovered that the reaction of BPA with phosgene (carbonyl chloride) produced a clear hard resin known as polycarbonate, which became widely used in the manufacture of plastics. It has been in commercial use since 1957.

Chemical structure of BPA

1.2. Uses of bisphenol A

Bisphenol A (2,2-bis (hydroxyphenyl) propane), an endocrine-disrupting chemical, is an environmental toxicant that has become an issue of controversy. Bisphenol A is one of the world's highest production volume chemicals (Ritter, 2011). It is used in polycarbonate plastics in many consumer products (e.g, water bottles, baby bottles) and epoxy resins lining food and beverage containers (EU, 2008; Kubwabo et al., 2009) and as a non–polymer additive to other plastics (Chitra et al., 2002; Hernandez- Rodriguez et al., 2007). It is also a key ingredient in the manufacture of epoxy resins that are used as dental sealants (Chapin et al., 2008) and is also found in polymers used in dental materials.

BPA containers

1.3. Bisphenol A exposure

The ubiquitous and extensive use of BPA-containing products results in a high human exposure worldwide (Vandenberg et al., 2010). The global population is subjected to repeated exposure to BPA, primarily through packaged food but also through drinking water, dental sealants, dermal exposure, and inhalation of household dusts (Lakind and Naiman, 2008) with detectable concentrations of metabolites in the urine of >90% of the population worldwide (Calafat et al., 2005; Ye et al., 2008). Evidence also indicates that exposure may occur through dermal contact with thermal papers used widely in cash register receipts (Biedermann et al., 2010). Heat, repeated washing of polycarbonate products and contact with either acidic or basic compounds accelerate the hydrolysis of the ester bond linking BPA molecules in polycarbonate plastics and resins leading to an increase in the rate of leaching of BPA (Welshons et al., 2006; Lim et al., 2009). BPA from polycarbonated bottles leAChEs into water at rates ranging from 0.2 to 0.79 ng/h and boiling water increases these rates up to 55 folds (Le et al., 2008). Hence, it becomes an integrated part of the food chain (Huang et al., 2012; Vandenberg et al., 2010). In addition, another potential source of human exposure to BPA is water used for drinking or bathing. Studies conducted in Japan (Kawagoshi et al., 2003) and in the United States (Coors et al., 2003) have shown that BPA accounts for most of the estrogenic activity that leAChEs from landfills into the surrounding ecosystem.

The World Health Organization (WHO) estimated that the mean BPA exposure for adults ranged from less than 0.01 to 0.40 lg/kg bw/day (95th percentile 0.06–1.5 lg/kg bw/day) and for young children to teenagers from 0.1–0.5 lg/kg bw/day (95th percentile 0.3–1.1 lg/kg bw/day) (WHO, 2011). Health Canada recently estimated that the general population has a mean BPA exposure of 0.055 lg/kg bw/day and that children between 0–18 months had exposures ranging from 0.083 to 0.164 lg/kg bw/day (Health Canada, 2012).

1.4. Current opinion of scientific organizations

To date, there is a controversy about the toxicity of BPA. Although FDA has labeled BPA as a safe agent (FDA, 2008), newly emerging data has raised demands for more studies to evaluate human health risk assessment of BPA exposure (Hugo et al., 2008). This is especially urgent in developing countries where plastic usage has increased exponentially and in certain population groups especially those suffering from malnutrition leading them to be at a higher risk than other populations (Aslan et al., 2006; Lahera et al., 2006). In 2010, the FDA indicated some concern about the safety of BPA and declared that more research is needed (Erickson, 2010). Health Canada (2008) has also banned the use of BPA from baby bottles. Similarly, the American Endocrine Society recommended further research on endocrine-disrupting chemicals such as BPA, raising a strong basis for concern about possible links between these chemicals, obesity, and related disorders (Diamanti-Kandarakis et al., 2010). In early 2011, the European Commission issued a new Directive setting a temporary ban on the use of BPA in the manufacture of polycarbonate infant feeding bottles in Europe (European Commission, 2011). The Danish Veterinary and Food Administration banned BPA in food containers and products for very young children in 2010. The Australian government has instituted a voluntary phase-out of polycarbonate baby bottles (FSANZ, 2012). Moreover, France has a ban on the use of BPA in baby bottles, and has introduced legislation in 2013 that would further ban the use of BPA for materials in contact with food for children between 0–3 years of age.

1.5. Mechanism of Action

Both in vivo and in vitro experiments demonstrated that BPA acts as an endocrine disrupting chemical (vom Saal and Hughes, 2005). The actions of BPA are mediated by endocrine-signaling pathways that act as powerful amplifiers, with the result that large changes in cell function can occur in response to extremely low concentrations (Welshons et al., 2003). BPA is an estrogen agonist that binds to both classical nuclear receptors [estrogen receptor α (ERα) and ERβ]. It has been suggested that exposure to low doses of BPA might interfere with normal estrogenic signaling (Welshons et al., 2006). BPA may not only act as an estrogen at doses within the range of human exposure but may also interfere with androgens, thyroid hormones, and cell signaling pathways (Wetherill et al. 2007).

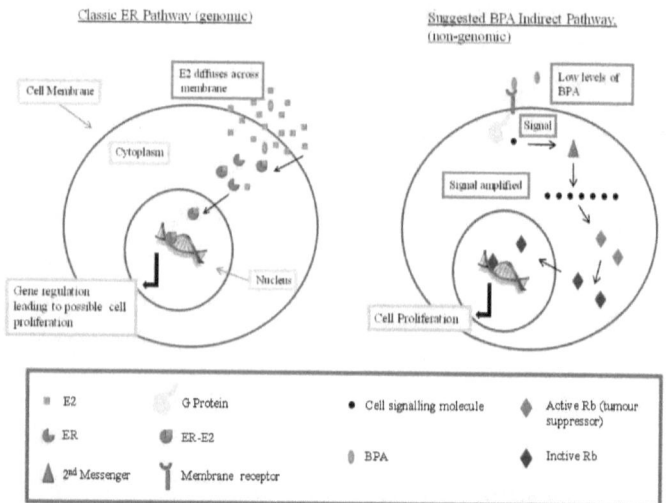

http://healthandenvironmentonline.com/2009/07/14/indirect-bpa-mechanism

1.6. Pharmacokinetics and metabolism

Atkinson and Roy (1995) reported that BPA is oxidized to a reactive metabolite 4, 5-bisphenol-O-quinone. BPA has been shown to decompose to many kinds of metabolites probably including BPA radical by a reaction of radical oxygen (Sajiki, 2001).

Several studies reported that BPA is metabolized mainly through conjugation to form the glucuronide metabolite (BPA glucuronide), which lacks estrogenic activity (Matthews et al., 2001; Calafat, 2011). Humans and laboratory animals form the glucuronide as the primary metabolite. Adult rats or mice rapidly metabolize orally-administered BPA to the glucuronide, which is then eliminated primarily via the bile (Doerge et al., 2010a). Non human primates and humans also rapidly metabolize orally-administered BPA to the glucuronide; however, it is eliminated primarily via the urine (Doerge et al., 2010b).

1.7. Health Hazards of Bisphenol A

In humans, increased levels of BPA in adults have been correlated with various diseases, health outcomes and medical conditions. Lang et al. (2008) reported a significant relationship between urine concentration of BPA and cardiovascular disorders, type 2 diabetes and liver enzyme abnormalities in a representative sample of US population. Moreover, two studies on laboratory animals have shown adverse effects of BPA on brain, reproductive system, metabolic processes, including alterations in insulin homeostasis and liver enzymes (Richter et al., 2007; Lang et al., 2008). In addition, absorption of large amounts of BPA through skin has been shown to cause extensive damage to liver, kidney and other vital organs in human (Sax, 1985; Suarez et al., 2000).

1.7.1. Effects on metabolism

Previous studies suggested that higher BPA exposure is related to individual components of the metabolic syndrome including hypertension (Bae et al., 2012; Shankar and Teppala, 2012), diabetes mellitus (Lang et al., 2008; Shankar and Teppala, 2011; Silver et al., 2011), insulin resistance (Wang et al., 2012), and obesity (Ropero et al., 2008; Newbold et al., 2009; Carwile and Michels, 2011; Shankar et al., 2012). Contributing to the potential for altered metabolic homeostasis, BPA has been shown to alter glucose homeostasis, increase pancreatic insulin content and induce insulin resistance in adult male mice (Alonso-Magdalena et al., 2006).

Numerous studies in rodents have found that prenatal and early postnatal BPA exposure is associated with increased body weight and fat deposition (Howdeshell et al., 1999; Rubin et al., 2001; Akingbemi et al., 2004; Nikaido et al., 2004; Miyawaki et al., 2007; Patisaul and Bateman, 2008; Somm et al., 2009; Hiyama et al., 2011; Wei et al., 2011; Xu et al., 2011). However, some studies found no associations (Ryan and Vandenbergh, 2006; Newbold et al., 2007; Ryan et al., 2010) while others found decreased body weight (Nagel et al., 1997; Honma et al., 2002; Tyl et al., 2002; Negishi et al., 2003; Alonso-Magdalena et al., 2010; Nakamura et al., 2012). Although most studies have examined only perinatal exposure to BPA, Akingbemi et al. (2004) compared perinatal and chronic postnatal exposure with

BPA in male rats and reported that perinatal exposure at 2.4 µg/kg/day was associated with increased body weight in adulthood but postnatal exposure (from weaning to adulthood) was not. Of the rodent studies that have linked perinatal BPA exposure to postnatal weight gain and adiposity, many have found that the increase occurred only at low doses and was not apparent until the animals reached sexual maturity (Howdeshell et al., 1999; Rubin et al., 2001; Akingbemi et al., 2004; Miyawaki et al., 2007; Patisaul and Bateman, 2008; Wei et al., 2011).

A small number of studies have examined the association of BPA and obesity in humans. Cross-sectional analyses of the National Health and Nutrition Examination Survey (NHANES) showed that both children and adults with urinary BPA concentrations in the second, third and fourth quartiles had higher odds of obesity and larger waist circumference than those in the lowest quartile (Carwile and Michels, 2011; Shankar et al., 2012; Trasande et al., 2012). Similarly, a cross-sectional study in China reported that higher urinary BPA concentrations among adults were positively correlated with overweight, abdominal obesity, insulin resistance, and diabetes (Ning et al., 2011; Wang et al., 2012).

1.7.2. Effects on the reproductive system

The majority of studies on BPA have focused on their endocrine disrupting effects and their potential adverse effects on the developing reproductive system. Several studies suggested that BPA showed testicular toxicity in rats and mice (Tohei et al., 1999; Takahashi and Oishi, 2001). Accumulation of BPA in male reproductive organs have some clinical implications since exposure to low doses of BPA during fetal life was shown to decrease the efficiency of sperm production in the offspring of male mice (Chitra et al., 2003). In rat liver, BPA decreases the activity of the male–specific cytochrome P450 isoforms, and testosterone 2α and 6β hydroxylase (Hanioka et al., 1998). The reduced cytochrome P450 has been shown to provoke reactive oxygen species (ROS) and these in turn impair sperm function (Griveau et al., 1995).

1.7.3. Effects on vital organs

Several studies reported the occurrence of oxidative toxicity after BPA exposure in rats and mice (Chitra et al., 2003; Gong and Han, 2006). Moreover, numerous reports have shown that BPA induced oxidative stress in vital organs as the liver, kidney and testis (Bindhumol et al., 2003; Chitra et al., 2003; Kabuto et al., 2004; Mourad and Khadrawy, 2012). BPA has been shown to cause the formation of multinucleated giant cells in rat liver hepatocytes. It also causes degeneration of renal tubules in the kidney of rat and mice (National Toxicology Program, 1982; Nakagawa and Tayama, 2000). BPA semiquinone, a radical intermediate, was found to be involved in DNA adduct formation along with peroxidase and hydrogen peroxide in rat hepatocytes in vitro (Atkinson and Roy, 1995). Moreover, Bindhumol et al. (2003) suggested that low doses of BPA generate ROS by decreasing the activities of antioxidant enzymes and increasing lipid peroxidation thereby causing oxidative stress in the liver of rats.

1.7.4. Effects on the cardiovascular system

Melzer et al. (2010) postulated several mechanisms by which BPA might increase the risk of cardiovascular diseases (CVDs), including reduced nitric oxide (NO) bioavailability, altered vascular reactivity to endothelin-1, oxidative stress and inflammation.

There has been increasing interest in the concept that oxygen free radicals and NO play an important role in the pathogenesis of cardiovascular diseases (Das, 2000). ROS are generated within the heart from several cellular sources including cardiac myocytes, endothelial cells, and neutrophils (Tsutsui et al., 2009). The heart has the highest oxygen uptake rate within the human body, consuming about 0.1 mL O_2/g per minute at basal rates (Antoni, 1991). To meet the demand for ATP synthesis by oxidative metabolism, cardiac myocytes have the highest density of mitochondria in the entire body. Under physiological conditions, small quantities of ROS are formed during mitochondrial respiration, which can, however, be eliminated by the endogenous antioxidant mechanisms of the myocytes.

However, when the production of ROS exceeds the capacity of the antioxidant defenses, oxidative stress might have a harmful effect on the functional and structural integrity of biological tissue. ROS result in contractile failure and structural damage in the myocardium (Tsutsui et al., 2009) possibly through the oxidation of proteins, DNA, and membrane phospholipids (McCord, 1985).

Nitric oxide as a gaseous free radical may act as a neurotransmitter and an effective cardiovascular modulator. This gas plays a fundamental role in cardiovascular physiology and pathophysiology (Shah et al., 1999). Within the cardiovascular system, NO participates in the regulation of coronary blood flow and tension of vessel wall (Roy et al., 2005). It is known to be deeply involved in vascular pathologies, such as hypertension (Moncada et al., 1991; Ignarro et al., 1999). It is hypothesized that chronic NO-deficient hypertension and alteration in heart rate are associated with depletion of antioxidants and oxidative injury to the heart (Husain and Hazelrigg, 2002). In addition, the deficiency in endothelial nitric oxide synthase (eNOS) has been shown to promote the development of heart failure postmyocardial infarction (Scherrer-Crosbie et al., 2001).

Acetylcholinesterase (AChE) is an important component of the heart's cholinergic system; it is known to regulate the cardiac parasympathetic responses by controlling acetylcholine levels

(Hoover et al., 2004). Normally, AChE rapidly and efficiently degrades acetylcholine, thereby terminating its signaling action (Lefkowitz et al., 1996).

Several investigators found that higher BPA concentrations were associated with cardiovascular diagnoses (Lang et al., 2008; vom Saal and Myers, 2008; Melzer et al., 2010) and incident coronary artery disease (Melzer et al., 2012). In addition, Asano et al. (2010) reported that BPA in the micromolar range activates Maxi-K (KCa1.1) ion channels in human coronary smooth muscle cells in culture to a degree sufficient to hyperpolarize the membrane potential. Pant et al. (2011) reported that acute exposure to BPA depressed cardiac activity even up to the stage of asystole. They suggested that the decreased contractility may lead to coronary insufficiency. Recently, several studies linked the exposure to BPA with atherosclerosis (Lind and Lind, 2011) and atherosclerosis development (Olsén et al., 2012).

1.7.5. Effects on the central nervous system

As BPA is lipophilic, it penetrates the blood–brain barrier (Sun et al., 2002) and can easily accumulate in the brain (Choi et al., 2007). Growing evidence suggests that BPA acts directly on neuronal functions within the central nervous system (CNS), where it modifies the activity of neuronal pathways and/or centers involved in nociception and pain (Choi et al., 2007).

Recently, several investigations were carried out to explore the neurotoxic effects of BPA. They showed that BPA induces persistent aberrations in spontaneous behavior and in learning and memory in rodents (Yu et al., 2011). Moreover, it has been reported that BPA causes synaptic remodeling in the nervous system and affects the development of higher cognitive functions (Hajszan and Leranth, 2010). It has also been reported to induce hippocampal neuronal apoptosis by increasing oxidative stress and altering MARK signaling (Lee et al., 2008). In an examination using in vivo microdialysis, Obata and Kubota (2000) found that BPA increased hydroxyl radical formation in the rat striatum. It has been suggested that pathological conditions induced by BPA may be causally related to the overproduction of ROS and free radicals generated by BPA metabolism and/or ER-mediated systems (Kabuto et al., 2003). In addition, Kim et al. (2011) showed that BPA has a neurotoxic effect on hippocampal neurogenesis and causes behavioral deficits.

Nakamura et al. (2010) reported that BPA disrupts monoamine neurotransmitter levels in some areas of adult mouse brain, thereby inducing morphological and molecular changes. Miyagawa et al. (2007) suggested that prenatal and neonatal exposures to BPA could induce other behavioral abnormalities associated with alteration of not only the dopaminergic system but also other neurotransmitters. The authors found that prenatal and neonatal exposure not only to high dose (2 mg/g diet) but also to low dose (30 ng/g diet) of BPA dramatically decreased cholinergic transmission in adult brain, resulting in learning and memory deficits. Several studies suggested that some neurotransmitter systems such as somatostatin and γ-aminobutyric acid (GABA) are linked to the predominant estrogenic effects of BPA in developing brain (Facciolo et al., 2005; Choi et al., 2007).

2. Current research on the effect of bisphenol A in adult rats

Humans are widely exposed to endocrine-disrupting chemicals, including BPA which is probably accumulated in lifetime. The widespread consumption of BPA-containing products has raised concerns among scientists and regulatory agencies that human exposure to BPA may have adverse effects on different vital organs. Most of the previous research has concentrated on the effects of the prenatal and/or postnatal exposure to BPA on different organs. Consequently, an extensive research was conducted in our laboratory to investigate the effects of BPA on different vital organs including the liver, testis, heart and brain in adult male albino rats (Mourad and Khadrawy, 2012; Aboul Ezz et al., 2015, Khadrawy et al, 2015).

2.1. Experimental Design

Animals were divided randomly into 4 groups. Group (1) served as control and received an oral administration of distilled water, five times a week, throughout the experimental protocol. They were divided to two subgroups which received distilled water for 6 weeks and 10 weeks. In Group (2), rats were administered orally with 25 mg/kg of BPA daily for 6 weeks. Animals of groups (3) and (4) received an oral administration of 10 mg/kg of BPA for 6 and 10 weeks, respectively. The doses of BPA were administered five times a week. The higher dose of BPA (25 mg/kg) in this study was based on previous studies (Bian et al. 2006; Richter et al. 2007). Each group was divided into two subgroups, one for the analysis of brain parameters and the other for the study of serum and other organs. A group of the control animals was sacrificed simultaneously with each group of the treated animals.

In the first subgroup of animals, the brain of each animal was transferred rapidly to an ice-cold petri dish where it was dissected to remove the hippocampus and cortex. The brain samples were divided into two equal halves, weighed and kept at -58°C until analyzed. The left half of each brain sample was homogenized in 5% w/v 20 mM phosphate buffer, pH 7.6, centrifuged, and used for the analysis of AChE activity and the levels of malondialdehyde (MDA) as a measure of lipid peroxidation, reduced glutathione (GSH) and nitric oxide (NO). The right half of each brain area was homogenized in 75% ethyl alcohol and used for the determination of amino acid neurotransmitters.

The liver, kidney, testes and heart of each animal of the second subgroup were quickly removed, homogenized in ice cold phosphate buffer (50 mM pH 7.4, 0.1 % triton X and 0.5 mM EDTA) and used for the analysis of different oxidative stress parameters. Blood samples from these animals were drawn from the retro-orbital venous plexus according to the method of Sorg and Buckner (1964). They were left to coagulate at room temperature, then centrifuged at 986 g for 15 minutes, the clear non-haemolyzed supernatant serum was quickly removed and used for the estimation of liver and kidney functions.

3. Discussion

3.1. The effect of BPA on the liver

In our study, we found a significant increase in lipid peroxidation and glutathione-S-transferase (GST) activity which was accompanied by a significant decrease in GSH in the liver of rats treated with BPA (25 mg/kg) for 6 weeks reflecting a state of oxidative stress in the hepatocytes (Mourad and Khadrawy, 2012).

Reduced glutathione (GSH) is the most important freely available antioxidant, which acts directly as an antioxidant and also participates in catalytic cycles of several antioxidant enzymes such as glutathione peroxidase, glutathione reductase and GST (Circu and Aw, 2008; Biswas and Rahman, 2009). In addition, GST catalyzes the conjugation of GSH – via a sulfhydryl group – to electrophilic centers on a wide variety of substrates (Douglas, 1987). This activity detoxifies endogenous compounds such as peroxidized lipids (Leaver and George, 1998) and catalyzes the breakdown of xenobiotics.

Mourad and Khadrawy (2012) concluded that the oxidative stress induced by BPA in the liver of rats treated with 25 mg/kg of BPA may be due to the generation of ROS and the formation of quinine radical, one of the BPA metabolites. They suggested that the raised activity of GST may be at the expense of the content of GSH that acts as a catalyst for GST. At the same time, GST catalyzes the conjugation of GSH to electrophilic centers of endogenous compounds to detoxify peroxidized lipids that recorded a significant increase in their study. This could explain the decreased content of GSH and the increased activity of GST that were recorded in the liver after the high dose.

In the liver of rats treated with BPA at a dose level of 10 mg/kg for 10 weeks, our team observed a significant decrease in GSH content and catalase activity and a significant increase in GST activity.

Catalase is a common enzyme found in nearly all living organisms that are exposed to oxygen, where it catalyzes the decomposition of hydrogen peroxide to water and oxygen. Bindhumol et al. (2003) suggested that the reduction in the activity of catalase may reflect the inability of liver mitochondria and microsomes to eliminate hydrogen peroxide produced after exposure to BPA. This may be due to the enzyme inactivation caused by excess ROS production in mitochondria (Pigeolet et al., 1990).

Mourad and Khadrawy (2012) suggested that the reduced activity of catalase after 10 weeks in the liver of rats treated with 10 mg/kg of BPA which was accompanied by reduced GSH content and increased GST activity could explain the ability of these antioxidant mechanisms to prevent the increase in lipid peroxidation and maintain its normal level at this time interval. On the other hand, the same authors found a significant increase in both aspartate aminotransferase (AST) and alanine aminotransferase (ALT) over control values in rats treated daily with 25 mg/kg for 6 weeks. However, the lower dose (10 mg/kg) did not affect the activities of both enzymes that reflect liver function after the two studied time intervals. Similarly, Korkmaz et al. (2010) reported a significant increase in ALT and AST activities in rats treated with 25 mg/kg BPA for 50 days.

Aspartate aminotransferase (AST) and alanine aminotransferase (ALT) are two enzymes of the most reliable markers of hepatocellular injury or necrosis. Their levels are elevated in a variety of hepatic disorders. Of the two, ALT is thought to be more specific for hepatic injury because it is present mainly in liver cytosol and in low concentration elsewhere (Giboney, 2005). When the liver hepatocytes are damaged, both enzymes are released into the blood where the significant increase in their activities indicates damage to the cytosol and also to the mitochondria (Mathuria and Verma, 2008). Mourad and Khadrawy (2012) suggested that the oxidative stress induced by the high dose of BPA (25 mg/kg/day for 6 weeks) may mediate the disturbance in hepatic function which is reflected by the increase in ALT and AST. The absence of any effect on hepatic function after the lower dose (10 mg/kg for 6 and 10 weeks) which was accompanied by normal lipid peroxidation may support this explanation.

3.2. The effect of BPA on the testis

In the testis, the study from our laboratory revealed a significant increase in lipid peroxidation that was accompanied by a significant decrease in GSH content and catalase activity after 6 weeks of daily administration of both the high (25 mg/kg) and low dose (10 mg/kg) of BPA (Mourad and Khadrawy, 2012). The authors suggested that these effects reflect a state of oxidative stress which may underlie BPA-induced testicular toxicity in rat and mice (Tohei et al., 1999; Takahashi and Oishi, 2001) and may be due to the fact that the testicular membranes are rich in polyenoic fatty acids that are prone to undergo peroxidative decomposition (Rosenblum et al., 1989).

3.3. The effect of BPA on the kidney

The data of Mourad and Khadrawy (2012) showed non significant changes in oxidative stress parameters in the kidney due to BPA treatment. However, serum uric acid increased significantly after 6 weeks of the daily oral administration of BPA at both the high (25 mg/kg) and low (10 mg/kg) dose levels.

It has been demonstrated that serum uric acid levels are significantly elevated in non alcoholic fatty liver patients (Li et al., 2009). Marmugi et al. (2011) suggested that low doses of BPA may influence de novo fatty acid synthesis thereby contributing to hepatic steatosis. Furthermore, uric acid increase may be due to the effect of BPA on the heart as several studies showed an association between elevated uric acid levels and cardiovascular diseases (Culleton et al., 1999; Fang and Alderman, 2000). In the light of the previous reports, Mourad and Khadrawy (2012) concluded that the increase in uric acid could not be attributed to impairment in kidney function but rather to effects on the heart and liver.

3.4. The effect of BPA on the heart

A recent study by our group revealed that BPA administration induced a state of oxidative stress in the heart of rats as evident from the increase in MDA levels and decrease in catalase activity at the two tested doses (10 and 25 mg/kg) after 6 weeks and the decrease in GSH levels after the administration of the two doses of BPA at all the tested time segments (Aboul Ezz et al., 2015). The authors suggested that BPA induced excessive ROS generation leading to increased lipid peroxidation and significantly compromised mitochondrial function.

As catalase converts hydrogen peroxide into hydrogen oxide, Aboul Ezz et al. (2015) suggested that the reduction in the activity of catalase may be due to the exhaustion of the enzyme in attempting to eliminate the hydrogen peroxide produced after the exposure to BPA.

Glutathione provides a first line of defense against ROS, as it can scavenge free radicals and reduce hydrogen peroxide (Pastorea et al., 2003). Aboul Ezz et al. (2015) reported a significant decrease in GSH levels after BPA administration at the two dose levels (25 mg/kg and 10 mg/kg) for different time intervals. The authors concluded that GSH was consumed during the conversion of hydrogen peroxide into hydrogen oxide. The peroxide was readily converted to the hydroxyl radical which may be involved in the observed decrease in GSH levels as GSH itself is also a general hydroxy-radical scavenger. Their findings were supported by various studies demonstrating that glutathione is reduced in the tissues by oxidative stress (Sian et al., 1994; Melchiorri et al., 1996).

Glutathione-S-transferase catalyses the conjugation of GSH with several compounds produced in vivo during oxidative stress. In our study, a significant increase in GST activity occurred at the high dose level of BPA (25 mg/kg for 6 weeks). This may eventually lead to the consumption of GSH during the generation of glutathione-S-conjugates by glutathione-S-transferases thus lowering the level of total intracellular glutathione after prolonged treatment.

Our data in the heart revealed a significant decrease in NO levels in rats receiving a daily oral administration of 25 mg/kg of BPA for 6 weeks and 10 mg/kg for 10 weeks (Aboul Ezz et al., 2015).

Nitric oxide is a free radical synthesized by the nitric oxide synthase (NOS) (Cannon, 1998). All three NOS isoforms such as constitutive (nNOS and eNOS) and inducible (iNOS) are expressed in the cardiovascular system (Kelly et al., 1996). It has been reported that normal endothelial release of NO through endothelium-derived NO (eNOS) reaction mediates physiologic vasodilation, while excessive release through iNOS induction may play a role in regulating blood pressure (BP), heart rate (HR) and endogenous antioxidants in septic shock (Petros et al., 1991). Acute or chronic administration of NOS inhibitors was reported to cause BP elevation and changes in HR in normal rats (Tribulová et al., 2000).

We postulated that the decreased level of NO under the effect of BPA in the heart may result in vasoconstriction which in turn may lead to decreased blood supply to the cardiac tissue. This may eventually cause a state of myocardial ischemia and consequently oxidative stress. Moreover, the decreased NO availability may represent one of the most important mechanisms mediating the reported CVDs related to BPA administration.

In our study, a significant decrease in the activity of AChE enzyme was observed in the rat heart after the daily oral administration of 25 mg/kg of BPA for 6 weeks and also after the administration of 10 mg/kg of BPA for 6 and 10 weeks (Aboul Ezz et al., 2015).

Acetylcholinesterase hydrolyzes acetylcholine and thereby terminates the action of this neurotransmitter at the cholinergic neuroeffector junctions of the heart. Few evidences suggest that inhibition of AChE is mediated by oxidative stress (Wyse et al., 2004). This was supported by the notion that hydroxyl radicals are involved in AChE inhibition (Tsakiris et al., 2000). Thus, the present inhibition of AChE activity may be related to the state of oxidative stress induced by BPA in rat heart.

Pant et al. (2011) found that BPA decreased the rate and force of atrial contractions simultaneously and depressed the functioning of the pacemaker cells and the contractile machinery of the heart. The authors suggested that the decreased rate and force of contractions can be due to the activation of cholinergic system or NO (Deshpande et al., 2008; Kanoo et al., 2009). From the data obtained in our laboratory, it may be postulated that the deficiency in NO levels together with the increased cholinergic activation resulting from AChE enzyme inhibition induced by BPA administration may lead to the previously reported reduction in the rate and force of cardiac contractions.

There are considerable data linking oxidative stress and ROS to the physiology and pathophysiology of CVD (Sugamura and Keaney, 2011). In addition, elevated levels of oxidative stress markers are detected in several pathologic conditions of cardiovascular disorders, including hypertension, ventricular hypertrophy, atherosclerosis, and congestive heart failure (Carlos et al., 1998; Keith et al., 1998; Miller et al., 1998; Harjai, 1999).

Both experimental and clinical studies suggested that the generation of ROS increases in heart failure (Hill and Singal, 1996, 1997; Mallat et al., 1998). Levels of lipid peroxides and 8-iso-prostaglandin F2a, the major biochemical markers of ROS generation, were found to be elevated in the plasma and pericardial fluid of patients with heart failure and were also positively correlated to its severity (Mallat et al., 1998).

On the other hand, depletion of GSH and GSH/GSSG ratio in blood has been reported to be a good marker in hypertension (Vaziri et al., 2000; Husain, 2002). It is clear that ROS may contribute to myocyte injury resulting from ischemic–reperfusion (Zweier et al., 1987), reduction of endogenous antioxidants in the myocardium (Hill and Singal, 1997), and the remodeling response (Dhalla et al., 1996).

Several reports found a link between urinary BPA concentrations and prevalence of heart diseases using 2003–2006 NHANES data, suggesting an association between BPA exposure and CVD (Lang et al., 2008; Melzer et al., 2010).

Aboul Ezz et al. (2015) concluded that the increase in lipid peroxidation and the reduction in the antioxidant mechanisms, NO level and AChE activity in the heart after the daily oral administration of BPA at the selected dose levels may lead to the generation of ROS and the development of a state of oxidative stress which may underlie the CVDs linked to BPA exposure.

3.5. The effect of BPA on body weight gain

Our data revealed an increase in the body weight gain of animals treated with the two doses of BPA (25 mg/kg and 10 mg/kg) for different time intervals (Aboul Ezz et al., 2015). Our results are consistent with those of other investigators who showed that perinatal BPA exposure increased body weight relative to controls (Rubin et al., 2001; Miyawaki et al., 2007).

The increase in body weight gain emphasizes the ability of BPA to promote obesity which in turn could exacerbate many of the metabolic and cardiovascular disorders reported after BPA exposure. This effect of BPA could be explained by the report of Hugo et al. (2008) who showed the ability of low levels of BPA to decrease adiponectin release from human adipose tissue explants. Adiponectin is known to play a positive role in cardiovascular health. Another possible explanation of enhanced weight gain in BPA-exposed animals is an increase in food intake by the estrogenic action of BPA on neuronal circuits in the hypothalamus that control appetite (Wade and Schneider, 1992). In view of the above data, BPA exposure could be a major public health concern in relation to the epidemic of childhood and adult obesity (Laron 2004; Reilly 2005).

3.6. The effect of BPA on the nervous system

The data from our laboratory revealed changes in cortical and hippocampal amino acid neurotransmitters after BPA administration which were accompanied by a state of oxidative stress. This was indicated from the significant increase in lipid peroxidation and NO level in the cortex after 6 weeks of exposure to BPA (25 mg/kg) and lipid peroxidation in the hippocampus after exposure to BPA (Khadrawy et al., 2015). The significant increase in cortical and hippocampal GSH level in our study may represent a compensatory mechanism to mitigate the state of oxidative stress induced by BPA, as GSH is the most important freely available antioxidant in the nervous system (Circu and Aw, 2008). We observed no significant changes in the cortical oxidative stress markers after the lowest dose of BPA (10 mg/kg for 6 and 10 weeks).

The brain is particularly vulnerable to oxidative damage, due to the high concentration of unsaturated lipids as well as the high rate of oxidative metabolism (Dringen et al., 2005). It has been suggested that BPA can cause oxidative stress that generates highly reactive membrane toxic intermediates in the brain (Aydoğan et al., 2008). Reactive oxygen metabolites such as hydroxyl radicals, peroxide anions, peroxyl radicals, and hydrogen peroxide can cause significant oxidative damage by attacking biomolecules such as membrane lipids, DNA, and proteins (Kabuto at al., 2003). This may explain the significant increase in lipid peroxidation induced by BPA in our present study.

The significant increase in the excitatory amino acids following exposure to BPA (25 mg/kg for 6 weeks), which was accompanied by a significant decrease in the inhibitory amino acids may result in cortical excitotoxicity (Khadrawy et al., 2015). Both glutamate and aspartate stimulate N-methyl-D-aspartate (NMDA) receptors whose prolonged depolarization induces a massive influx of calcium ions (Ca^{2+}). Prolonged elevation of Ca^{2+} ions is believed to initiate a complex cascade of intracellular events that lead to neuronal destruction (Weber, 2012).

Khadrawy et al. (2015) suggested that the increase in the content of cortical and hippocampal excitatory amino acid neurotransmitters after BPA exposure could underlie the generation of ROS and inhibition of neuronal progenitor cells and finally cell death. Although the exposure of rats to BPA (10 mg/kg for 10 weeks) induced a significant increase in cortical glutamate and aspartate accompanied by a significant decrease in GABA and glycine, the authors found no changes in cortical oxidative stress markers. They suggested that these changes in amino acid neurotransmitters may represent the onset of effects and need to be followed up.

The hippocampus compared to other areas of the brain is characterized by a small extracellular space (Green et al., 1970). The limited extracellular space would result in a rapid buildup of extracellular concentrations of potassium ions observed during repetitive neuronal firing, inducing a reduction in neuronal membrane potential, and increasing neuronal excitability (Fisher et al., 1976). This may explain the rapid effects in amino acid neurotransmitters and oxidative stress markers obtained in the hippocampus after 6 weeks of exposure to BPA (10mg/kg) in the current study. However, in the hippocampus, Khadrawy et al. (2015) found that the significant increases in the excitatory amino acid neurotransmitters were accompanied by significant increases in the inhibitory amino acids GABA and glycine. The authors suggested that this may reflect an attempt to counteract the increased excitability in this particular area. The recovery of the amino acid neurotransmitter levels after 10 weeks of BPA (10 mg/kg) supported their suggestion. However, this attempt did not prevent the damaging effects of the generated free radicals as evident from the increased lipid peroxidation levels obtained after 10 weeks of exposure to BPA (10 mg/kg). Supporting our findings is the study of Tanabe et al. (2012) who stated that BPA may reach the brain and accumulate without detoxification as judged from no significant conversion of 3H-BPA to other metabolites. Our results are consistent with a significant level of BPA accumulated in the hippocampus of adult male rats.

N-methyl-D-aspartate (NMDA) receptors, a subtype of glutamate receptor, play a crucial role for normal brain functions. BPA markedly inhibited the expressions of NMDA subunits in the hippocampus of male offspring during postnatal developmental stage and adulthood (Xu et al., 2010). This reported inhibition in the expression of NMDA receptors may be a strategy by which the brain attempts to minimize the state of excitotoxicity induced by BPA as evident from the data of Khadrawy et al. (2015).

BPA has been implicated as an endocrine-disrupting chemical due to its ability to mimic the action of endogenous estrogenic hormones (Paris et al., 2002). It has been reported that estradiol increased the concentration of glutamate in the arculate nucleus (Blutstein et al., 2009). It could also modulate glutamatergic neurotransmission in various brain regions (Yokomaku et al., 2003) where it increases excitability. Although the affinity of ERα for BPA is very week (Kuiper et al., 1997), estrogen-related receptor γ is a high-affinity binding protein for BPA (Takayanagi et al., 2006). It has been reported that blocking NMDA receptors by MK-801 abolishes the enhancing effects of estradiol and BPA (Ogiue-Ikeda et al., 2008). Khadrawy et al. (2015) suggested that the increase in the excitatory amino acids in the hippocampus and cortex could be attributed to an endocrine mechanism through the action of BPA on estrogen and estrogen-related receptors (estradiol-like effect).

It has been found that cholinergic fibers were dramatically decreased in several hippocampal regions of mice prenatally and neonatally exposed to low and high doses of BPA (Miyagawa et al., 2007). The authors suggested that chronic exposure to BPA could induce memory impairment through the reduction in acetylcholine production in the hippocampus. The decrease in acetylcholine could also be attributed to the elevated activity of AChE recorded by our team as AChE is the enzyme responsible for the breakdown of acetylcholine. Viberg and Lee (2012) showed that adult mice neonatally exposed to BPA exhibited an abnormal spontaneous behavior to a novel home environment manifested as reduced habituation and hyperactive condition. The behavioral disturbances were long lasting and irreversible. They also suggested that neonatally exposed mice are incapable of processing new sensory information in a novel home environment and integrate it to a normal habitation capacity indicating reduced cognitive capacity.

Khadrawy et al. (2015) concluded that the prominent neurochemical effects of BPA in the cortex and hippocampus that were evident from the significant changes in amino acid and cholinergic neurotransmission associated with oxidative stress in the present study may be attributed to the deficiency of detoxification mechanisms for BPA in the brain (Doerge et al., 2010a, 2011). It has been reported that low-dose exposure to BPA shows rather weak toxic effects on the reproductive or endocrine functions in the peripheral tissues, probably due to the efficient detoxification of BPA by the liver. However, low dose exposure to BPA may significantly affect brain functions because the detoxification of BPA in the brain is probably very weak due to the extremely low expression of drug-metabolizing enzymes in the brain (Kishimoto et al., 2004; Chinta et al., 2005).

The present changes induced by BPA could underlie the disturbances in many neuronal functions such as memory, cognitive impairments, hyperactivity, and the inhibition of neuronal progenitor cells proliferation. In the present study, the similarity between the effects of 25 mg/kg for 6 weeks with 10 mg/kg for 10 weeks indicates that BPA has accumulating effects. This in turn may raise the level of warning from the exposure to low levels of BPA during lifetime.

4. Conclusion

In the light of the present studies, it could be concluded that high doses of BPA have serious effects on the liver, testis, heart and brain. These effects are mediated by the oxidative stress induced by BPA. Although low doses of BPA induced minor changes in oxidative stress parameters after 6 weeks in several tissues, significant changes were obtained after 10 weeks suggesting that low doses may produce drastic effects after long term exposure. In addition, it is evident that BPA has a cardiotoxic effect which is mediated not only by the generation of ROS and reduction of antioxidant defenses of the heart which aggravate a state of oxidative stress but also by the concomitant reduction in NO levels and AChE activity. Moreover, the increase in body weight may contribute to the cardiovascular disturbances resulting from BPA. It is clear that these pathological conditions may occur after prolonged exposure to BPA even at extremely low levels. Furthermore, the results of the present studies suggest that in a population of high use of plastics where there is a great chance of exposure to BPA, the males may suffer from sexual disturbances due to the oxidative stress induced in the testis. A serious impact of BPA exposure on the brain of adult rats was also evident even at low doses due to BPA's accumulating effects. These data may raise concerns about the exposure of humans to BPA due to its wide applications in industry.

5. Recommendation

The widespread exposure to BPA necessitates further research on its long term effects in different organs and raises the demands for reconsidering the use of BPA in plastic industries.

Until the incorporation of BPA in different plasticizers and other industries is banned, its use should be limited and the erroneous handling of plastic containers should be avoided to reduce the health risks resulting from exposure to this endocrine disruptor.

How You Can Avoid Bisphenol A (BPA)

DO avoid plastic containers that have "PC" on the recycling label
DO avoid plastic containers with the number seven on the bottom
DO look for labels that say "BPA free"

DON'T use any plastic containers or bottles with scratches
DON'T heat up food or drinks in plastic containers
DON'T handle receipts from any store
DON'T use hand sanitizer after touching a receipt
(it increases the amount of BPA in your skin)
DON'T put plastic containers in the dishwasher

Learn More @DecodedScience.com

6. References

Aboul Ezz HS, Khadrawy Y A, Mourad I M (2015). The effect of bisphenol A on some oxidative stress parameters and acetylcholinesterase activity in the heart of male albino rats. Cytotechnology, 67: 145–155

Akingbemi BT, Sottas CM, Koulova AI, Klinefelter GR, Hardy MP (2004). Inhibition of testicular steroidogenesis by the xenoestrogen bisphenol A is associated with reduced pituitary luteinizing hormone secretion and decreased steroidogenic enzyme gene expression in rat Leydig cells. Endocrinology, 145: 592–603.

Alonso-Magdalena P, Morimoto S, Ripoll C, Fuentes E, Nadal A (2006). The estrogenic effect of bisphenol A disrupts pancreatic beta-cell function in vivo and induces insulin resistance. Environ Health Perspect, 114: 106–112.

Alonso-Magdalena P, Vieira E, Soriano S, Menes L, Burks D, Quesada I, Nadal A. (2010). Bisphenol A exposure during pregnancy disrupts glucose homeostasis in mothers and adult male offspring. Environ Health Perspect, 118: 1243–1250.

Antoni H (1991). Function of the heart, in: Schmidt, R.F., Thews, G. (Eds), Human Physiology. Berlin, Heidelberg, New York: Springer-Verlag, pp. 358–396.

Asano S, Tune JD, Dick GM (2010). Bisphenol A activates Maxi-K (K(Ca)1.1) channels in coronary smooth muscle. Br J Pharmacol, 160: 160–170.

Aslan M, Horoz M, Kocyigit A, Ozgonul S, Celik H, Celik M, Erel O (2006). Lymphocyte DNA damage and oxidative stress in patients with iron deficiency anaemia. Mutat Res, 601: 144-149.

Atkinson A, Roy D (1995). In vitro conversion of environmental estrogenic chemical bisphenol to DNA binding metabolite(s). Biochem Biophys Res Commun, 210: 424-433.

Aydoğan M, Korkmaz A, Barlas N, Kolankaya D (2008). The effect of vitamin C on bisphenol A, nonylphenol and octylphenol induced brain damages of male rats. Toxicology, 249: 35-39.

Bae S, Kim JH, Lim Y-H, Park HY, Hong Y-C (2012). Associations of bisphenol A exposure with heart rate variability and blood pressure. Hypertension, 60: 786-793.

Bian Q, Qian J, Xu L, Chen J, Song L, Wang X (2006). The toxic effect of 4-tert-octylphenol on the reproductive system of male rats. Food Chem Toxicol, 44: 1355-1361.

Biedermann S, Tschudin P, Grob K (2010). Transfer of bisphenol A from thermal printer paper to the skin. Anal Bioanal Chem, 398: 571-576.

Bindhumol V, Chitra KC, Mathur PP (2003). Bisphenol A induces reactive oxygen species generation in the liver of male rats. Toxicology, 188: 117-124.

Biswas SK, Rahman I (2009). Environmental toxicity, redox signaling and lung inflammation: The role of glutathione. Mol Aspects Med, 30: 60-76.

Blutstein T, Baab PJ, Zielke HR (2009). Hormonal modulation of amino acid neurotransmitter metabolism in the arcuate nucleus of the adult female rat: a novel action of estradiol. Endocrinology, 150: 3237-3244.

Calafat A (2011). Background Paper on BPA Biomonitoring and Biomarker Studies. World Health Organization, Geneva.

Calafat AM, Kuklenyik Z, Reidy JA, Caudill SP, Ekong J, Needham LL (2005). Urinary concentrations of bisphenol A and 4-nonylphenol in a human reference population. Environ Health Perspect, 113: 391-395.

Cannon RO (1998). Role of nitric oxide in cardiovascular disease: focus on the endothelium. Clin Chem, 44: 1809-1819.

Carlos DM, Goto S, Urata Y, Iida T, Cho S, Niwa M, Tsuji Y, Kondo T (1998). Nicardipine normalizes elevated levels of antioxidant activity in response to xanthine oxidase-induced oxidative stress in hypertensive rat heart. Free Radic Res, 29: 143-150.

Carwile JL, Michels KB (2011). Urinary bisphenol A and obesity: NHANES 2003–2006. Environmental Research, 111: 825–830.

Chapin RE, Adams J, Boekelheide K, Gray LE, Hayward SW, Lees PS, McIntyre BS, Portier KM, Schnorr TM, Selevan SG, Vandenbergh JG, Woskie SR (2008). NTP-CERHR expert panel report on the reproductive and developmental toxicity of bisphenol A. Birth Defects Res B: Dev Reprod Toxicol, 83: 157-395.

Chinta SJ, Kommaddi RP, Turman CM, Strobel HW, Ravindranath V (2005). Constitutive expression and localization of cytochrome P-450 1A1 in rat and human brain: presence of a splice variant form in human brain. J Neurochem, 93: 724-736.

Chitra KC, Latchoumycandane C, Mathur PP (2003). Induction of oxidative stress by bisphenol A in the epididymal sperm of rats. Toxicology, 185: 119–127.

Chitra KC, Latchoumycandane C, Mathur PP (2002). Effect of nonylphenol on the antioxidant system in epididymal sperm of rats. Arch Toxicol, 76: 545-551.

Chitra KC, Rao KR, Mathur P (2003). Effect of bisphenol A and co - administration of bisphenol A and vitamin C on epididymis of adult rats: a histopathological and biochemical study. Asian J Androl, 5: 203-208.

Choi IS, Cho JH, Park EJ, Park JW, Kim SH, Lee MG, Choi BJ, Jang IS (2007) . Multiple effects of bisphenol A, an endocrine disrupter, on GABA(A) receptors in acutely dissociated rat CA3 pyramidal neurons. Neurosci Res, 59: 8-17.

Circu ML, Aw TY (2008). Glutathione and apoptosis. Free Radic Res, 42: 689-706.

Coors A, Jones PD, Giesy JP, Ratte HT (2003). Removal of estrogenic activity from municipal waste landfill leachate assessed with a bioassay based on reporter gene expression. Environ Sci Technol, 37: 3430-3434.

Culleton BF, Larson MG, Kannel WB, Levy D (1999). Serum uric acid and risk for cardiovascular disease and death: The Framingham Heart Study. Ann Intern Med, 131: 7-13.

Das UN (2000). Free radicals, cytokines and nitric oxide in cardiac failure and myocardial infarction. Mol Cell Biochem, 215: 145-152.

Deshpande SB, Kanoo S, Alex AB (2008). Bradycardia induced by Mesobuthus tamulus scorpion venom involves muscarinic receptor-G-protein-coupled cell signaling pathways. Indian J Exp Biol, 46: 229-233.

Dhalla AK, Hill MF, Singal PK (1996). Role of oxidative stress in transition of hypertrophy to heart failure. J Am Coll Cardiol, 28: 506-514.

Diamanti-Kandarakis E, Palioura E, Kandarakis SA, Koutsilieris M (2010). The impact of endocrine disruptors on endocrine targets. Horm Metab Res, 42: 543-552.

Doerge DR, Twaddle NC, Vanlandingham M, Fisher JW (2010a). Pharmacokinetics of bisphenol A in neonatal and adult Sprague-Dawley rats. Toxicol Appl Pharmacol, 247: 158-165.

Doerge DR, Twaddle NC, Woodling KA, Fisher JW (2010b). Pharmacokinetics of bisphenol A in neonatal and adult rhesus monkeys. Toxicol Appl Pharmacol, 248: 1-11.

Doerge DR, Twaddle NC, Vanlandingham M, Brown RP, Fisher JW (2011). Distribution of bisphenol A into tissues of adult, neonatal, and fetal Sprague–Dawley rats. Toxicol Appl Pharmacol, 255: 261-270.

Douglas KT (1987). Mechanism of action of glutathione-dependent enzymes. Adv Enzymol Relat Areas Mol Biol, 59: 103-67.

Dringen R, Pawlowski PJ, and Hirrlinger J (2005). Peroxide detoxification by brain cells. J Neurosci Res, 79: 157-165.

Erickson B (2010). FDA raises flag on bisphenol A. Chem Eng News, 88: 8.

European Commission (2011). Commission Directive 2008/8/EU of 28 January 2011 amending Directive 2002/72/EC as regards to the restriction of the use of
Bisphenol A in plastic infant feeding bottles. Official Journal of the European Union 26, 11-14.

European Union (EU) (2008). Updated European Risk Assessment Report: 4,4- isopropylidenediphenol (bisphenol-A) (CAS Number: 80-05-7, EINECS Number: 201-245-8). Available from: http://ecb.jrc.it/documents/Existingchemicals/   RISK

ASSESSMENT/ADDENDUM/bisphenolaadd 325.pdf. Accessed 01.02.11.

Facciolo RM, Madeo M, Alò R (2005). Neurobiological effects of bisphenol A may be mediated by somatostatin subtype 3 receptors in some regions of the developing rat brain. Toxicol. Sci, 88: 477-484.

Fang J, Alderman MH (2000). Serum uric acid and cardiovascular mortality. The NHANES I epidemiologic follow-up study, 1971–1992. J Am Med Assoc, 283: 2404-2410.

FDA (2008). Draft assessment of bisphenol A for use in food contact applications. US Food and Drug Administration, Silver Spring, MD.

FDA (2010). Advancing regulatory science for public health. US Food and Drug Administration, Silver Spring, MD.

Fisher RS, Pedley TA, Moody WJ (1976). The role of extracellular potassium in hippocampal epilepsy. Arch Neurol, 33: 76-83.

FSANZ (2012). Regulation and monitoring of BPA. Food Standards Agency of Australia and New Zealand, Canberra, Australia. http://www.foodstandards.gov.au/consumerinformation/bisphenolabpa/regulationandmonitor5377.cfm.>

Giboney PT (2005). Mildly elevated liver transaminase levels in the asymptomatic patient. Am Fam Physician, 71: 1105-1110.

Gong Y, Han, XD (2006). Nonylphenol- induced oxidative stress and cytotoxicity in testicular sertoli cells. Reprod Toxicol, 22: 623-630.

Green JR, Halpern LM, Niel SV (1970). Alterations in the activity of selected enzymes in the chronic isolated cerebral cortex of cat. Brain, 93: 57-64.

Griveau JF, Dumont E, Renard P, Callergari JP, le Lannou D (1995). Reactive oxygen species, lipid peroxidation and enzymatic defense systems in human spermatozoa. J Reprod Fertil, 103:17-26.

Hajszan T, Leranth C (2010). Bisphenol A interferes with synaptic remodeling. Front Neuroendocrinol, 31: 519-530.

Hanioka N, Jinno H, Nishimura T, Ando, M (1998). Suppression of male- specific Research Article ISSN 2250-0480 Vol 2/Issue 2/Apr-Jun 2012 L-27 Life Science Zoology cytochrome P450 isoforms by bisphenol A in rat liver. Arch Toxicol, 72: 387-394.

Harjai KJ (1999). Potential new cardiovascular risk factors: left ventricular hypertrophy, homocysteine, lipoprotein(a), triglycerides, oxidative stress, and fibrinogen. Ann Intern Med, 131: 376-386.

Health Canada (2008). Government of Canada Takes Action on Another Chemical of Concern: Bisphenol A [Press Release]. Available: http://www.hc-sc.gc.ca/ahc-asc/media/nrcp/2008/2008_59-eng.php. Accessed 25 September 2009.

Health Canada (2012). Updated Assessment of Bisphenol A (BPA) Exposure from Food Sources Health Canada, Bureau of Chemical Safety, Ottawa, Ontario, Canada. <http://www.hc-sc.gc.ca/fn-an/alt_formats/pdf/securit/packag-emball/ bpa/bpa_hra-ers-2012-09-eng.pdf>.

Hernandez-Rodriguez G, Zumbado M, Luzardo OP, Monterde JG, Blanco A, Boada LD (2007). Multigenerational study of the hepatic effects exerted by the consumption of Haniokanonylphenol and 4-octylphenol contaminated drinking water in Sprague-Dawley rats. Environ Toxicol Pharmacol, 23: 73-81.

Hill MF, Singal PK (1996). Antioxidant and oxidative stress changes during heart failure subsequent to MI in rats. Am J Pathol, 148: 291-300.

Hill MF, Singal PK (1997). Right and left myocardial antioxidant responses during heart failure subsequent to myocardial infarction. Circulation, 96: 2414-2420.

Hiyama M, Choi EK, Wakitani S, Tachibana T, Khan H, Kusakabe KT, Kiso Y (2011). Bisphenol-A (BPA) affects reproductive formation across generations in mice. J Vet Med Sci, 73: 1211-1215.

Honma S, Suzuki A, Buchanan DL, Katsu Y, Watanabe H, Iguchi T (2002). Low dose effect of in utero exposure to bisphenol A and diethylstilbestrol on female mouse reproduction. Reprod Toxicol, 16: 117-122.

Howdeshell KL, Hotchkiss AK, Thayer KA, Vandenbergh JG, vom Saal FS (1999). Exposure to bisphenol A advances puberty. Nature, 401: 763-764.

Hoover DB, Ganote CE, Ferguson SM, Blakely RD, Parsons RL (2004). Localization of cholinergic innervation in guinea pig heart by immunohistochemistry for high-affinity choline transporters. Cardiovasc Res, 62: 112-121.

Huang YQ, Wong CK, Zheng JS, Bouwman H, Barra R, Wahlstrom B, Neretin L, Wong MH (2012). Bisphenol A (BPA) in China: A review of sources, environmental levels, and potential human health impacts. Environ Int, 42: 91-9.

Hugo ER, Brandebourg TD, Woo JG, Loftus J, Alexander JW, Ben-Jonathan N (2008). Bisphenol A at environmentally relevant doses inhibits adiponectin release from human adipose tissue explants and adipocytes. Environ Health Perspect, 116: 1642-1647.

Husain K (2002). Exercise conditioning attenuates the hypertensive effects of nitric oxide synthase inhibitor in rat. Mol Cell Biochem, 231: 129-137.

Husain K, Hazelrigg SR (2002). Oxidative injury due to chronic nitric oxide synthase inhibition in rat: effect of regular exercise on the heart. Biochim Biophys Acta, 1587: 75- 82.

Ignarro LJ, Cirino G, Casini A, Napoli C (1999). Nitric oxide as a signaling molecule in the vascular system: an overview. J Cardiovasc Pharmacol, 34: 879-886.

Kabuto H, Amakawa M, Shishibori T (2004). Exposure to bisphenol A during embryonic/fetal life and infancy increases oxidative injury and causes underdevelopment of the brain and testis in mice. Life Sci, 74: 2931-2940.

Kabuto H, Hasuike S, Minagawa N, Shishibori T (2003). Effects of bisphenol A on the metabolisms of active oxygen species in mouse tissues. Environ Res, 93: 31-35.

Kanoo S, Mandal MB, Alex AB (2009). Deshpande SB.Cardiac dysrhythmia produced by Mesobuthus tamulus venom involves NO-dependent G-Cyclase signaling pathway. Naunyn Schmiedebergs Arch Pharmacol, 379: 525-532.

Kawagoshi Y, Fujita Y, Kishi I, Fukunaga I (2003). Estrogenic chemicals and estrogenic activity in leachate from municipal waste landfill determined by yeast two-hybrid assay. J Environ Monit, 5: 269-274.

Keith M, Geranmayegan A, Sole MJ, Kurian R, Robinson A, Omran AS, Jeejeebhoy K N (1998). Increased oxidative stress in patients with congestive heart failure. J Am Coll Cardiol, 31: 1352-1356.

Kelly RA, Balligand JL, Smith TW (1996). Nitric oxide and cardiac function. Circ Res, 79: 363-380.

Khadrawy YA, Noor NA, Mourad IM, Ezz HS (2015). Neurochemical impact of bisphenol A in the hippocampus and cortex of adult male albino rats. Toxicol Ind Health, Apr 22. pii: 0748233715579803. [Epub ahead of print]

Kim ME, Park HR, Gong EJ, Choi SY, Kim HS, Lee J (2011). Exposure to bisphenol A appears to impair hippocampal neurogenesis and spatial learning and memory. Food Chem Toxicol, 49: 3383-3389.

Kishimoto W, Hiroi T, Shiraishi M, Osada M, Imaoka S, Kominami S, Igarashi T, Funae Y (2004). Cytochrome P450 2D catalyze steroid 21-hydroxylation in the brain. Endocrinology, 145: 699-705.

Korkmaz A, Aydogan M, Kolankaya D, Barlas N (2010). Influence of vitamin C on bisphenol A, nonylphenol and octyl -phenol induced oxidative damages in liver of male rats. Food Chem Toxicol, 48: 2865- 2871.

Kubwabo, C, Kosarac I, Stewart B, Gauthier BR, Lalonde K, Lalonde PJ (2009). Migration of bisphenol A from plastic baby bottles, baby bottle liners and reusable polycarbonate drinking bottles. Food Addit Contam Part A Chem Anal Control Expo Risk Assess, 26: 928-937.

Kuiper GG, Carlsson B, Grandien K, Enmark E, Häggblad J, Nilsson S, Gustafsson JA (1997). Comparison of the ligand binding specificity and transcript tissue distribution of estrogen receptors alpha and beta. Endocrinology, 138: 863-870.

Lahera V, Goicoechea M, de Vinuesa SG, Oubĩna P, Cachofeiro V, Gómez-Campderá F, Amann R, Luño J (2006). Oxidative stress in uremia: the role of anaemia correction. J Am Soc Nephrol, 12 (Suppl 3): S174-S177.

Lakind JS, Naiman DQ (2008). Daily intake of bisphenol A and potential sources of exposure: 2005–2006 National Health and Nutrition Examination Survey. J Expo Sci Environ Epidemiol, 21: 272-279.

Lang IA, Galloway TS, Scarlett A, Henley WE, Depledge M, Wallace RB, Melzer D (2008). Association of urinary bisphenol A concentration with medical disorders and laboratory abnormalities in adults. JAMA, 300: 1303–1310.

Laron Z (2004). Increasing incidence of childhood obesity. Pediatr Endocrinol Rev, 1(suppl 3): 443–447.

Le HH, Carlson EM, Chua JP, Belcher SM (2008). Bisphenol A is released from polycarbonate drinking bottles and mimics the neurotoxic actions of estrogen in developing cerebellar neurons. Toxicol Lett, 176: 149-156.

Leaver MJ, George SG (1998). A piscine glutathione S-transferase which efficiently conjugates the end-products of lipid peroxidation. Marine Environ Res, 46: 71-74.

Lee S, Suk K, Kim IK, , Jang IS, Park JW, Johnson VJ, Kwon TK, Choi BJ, Kim SH (2008). Signaling pathways of bisphenol A-induced apoptosis in hippocampal neuronal cells: role of calcium-induced reactive oxygen species, mitogen-activated protein kinases, and nuclear factor-kappa B. J Neurosci Res, 86: 2932-2942.

Lefkowitz RJ, Hoffman BB, Taylor P (1996). Neurotransmission: The autonomic and somatic motor nervous systems. In: Hardman JG; Limbird LL; Molinoff PB; Ruddon RW; Gilman AG (eds). Goodman and Gilman's The Pharmacological Basis of Therapeutics, 9th ed. New York, McGraw-Hill, pp 105-139.

Li Y, Xu C, Yu C, Xu L, Min M (2009). Association of serum uric acid level with nonalcoholic fatty liver disease: a cross-sectional study. J Hepatol, 50: 1029-1034.

Lim DS, Kwack SJ, Kim KB, Kim HS, Lee BM (2009). Potential risk of bisphenol A migration from polycarbonate containers after heating, boiling, and microwaving. J Toxicol Environ Health A, 72: 1285-1291.

Lind PM, Lind L (2011). Circulating levels of bisphenol A and phthalates are related to carotid atherosclerosis in the elderly. Atherosclerosis, 218: 207-213.

Mallat Z, Philip I, Lebret M, Chatel D, Maclouf J, Tedgui A (1998). Elevated levels of 8-iso-prostaglandin F2alpha in pericardial fluid of patients with heart failure: a potential role for in vivo oxidant stress in ventricular dilatation and progression to heart failure. Circulation, 97: 1536-1539.

Marmugi A, Ducheix,S., Lasserre,F., Polizzi, A., Paris,A., Priymenko,N., Bertrand-Michel, J., Pineau, T., Guillou, H., Martin, P.G. and Muselli- Lakhal, L (2011). Low doses of bisphenol A induce gene expression related to lipid synthesis and trigger triglyceride accumulation in adult mouse liver. Hepatology, 55: 395-407.

Matthews, J.B., Twomey, K., Zacharewski, T.R., 2001. In vitro and in vivo interactions of bisphenol A and its metabolite, bisphenol A glucuronide, with estrogen receptors alpha and beta. Chem. Res. Toxicol. 14, 149–157.

Mathuria N, Verma RJ (2008). Ameliorative effect of curcumin on aflatoxin-induced toxicity in serum of mice. Acta Pol Pharmaceut Drug Res, 65: 339-343.

McCord JM (1985). Oxygen-derived free radicals in postischemic tissue injury. N Engl J Med, 312: 159-163.

Melchiorri D, Reiter RJ, Sewerynek E, Hara M, Chen L, Nistico G (1996). Paraquat toxicity and oxidative damage. Reduction by melatonin. Biochem Pharmacol, 51: 1095-1099.

Melzer D, Osborne NJ, Henley WE, Cipelli R, Young A, Money C, McCormack P, Luben R, Khaw KT, Wareham NJ, Galloway TS (2012). Urinary bisphenol A concentration and risk of future coronary artery disease in apparently healthy men and women. Circulation, 125: 1482-1490.

Melzer D, Rice NE, Lewis C, Henley WE, Galloway TS (2010). Association of urinary bisphenol A concentration with heart disease: evidence from NHANES 2003/06. PLoS One, 5: e8673.

Miller FJJ, Gutterman DD, Rios CD, Heistad DD, Davidson BL (1998). Superoxide production in vascular smooth muscle contributes to oxidative stress and impaired relaxation in atherosclerosis. Circ Res, 82: 1298-1305.

Miyagawa K, Narita M, Narita M, Akama H, Suzuki T (2007). Memory impairment associated with a dysfunction of the hippocampal cholinergic system induced by prenatal and neonatal exposures to bisphenol-A. Neurosci Lett, 418: 236-241.

Miyawaki J, Sakayama K, Kato H, Yamamoto H, Masuno H (2007). Perinatal and postnatal exposure to bisphenol a increases adipose tissue mass and serum cholesterol level in mice. J Atherosclerosis Thrombosis, 14: 245–252.

Moncada S, Palmer RMJ, Higgs EA (1991). Nitric oxide: physiology, pathophysiology, and pharmacology. Pharmacol Rev, 43: 109-142.

Mourad IM, Khadrawy YA (2012). The sensitivity of liver, kidney and testis of rats to oxidative stress induced by different doses of bisphenol A. Int J Life Sci Pharma Res, 2: L19-L28.

Nagel SC, vom Saal FS, Thayer KA, Dhar MG, Boechler M, Welshons WV (1997). Relative binding affinity-serum modified access (RBA-SMA) assay predicts the relative in vivo bioactivity of the xenoestrogens bisphenol A and octylphenol. Environ Health Perspect, 105: 70-76.

Nakagawa, Y. and Tayama, S. Metabolism and cytotoxicity of bisphenol A and other bisphenols in isolated rat hepatocytes. Arch. Toxicol., 2000; 74: 99-105.

Nakamura K, Itoh K, Dai H, Han L, Wang X, Kato S, Sugimoto T, Fushiki S (2012). Prenatal and lactational exposure to low-doses of bisphenol A alters adult mice behavior. Brain Dev, 34: 57-63.

Nakamura K, Itoh K, Yoshimoto K, Sugimoto T, Fushiki S (2010). Prenatal and lactational exposure to low-doses of bisphenol A alters brain monoamine concentration in adult mice. Neurosci Lett, 484: 66-70.

National Toxicology Program (1982). Carcinogenesis bioassay of bisphenol A in F344 rats and B6C3F mice. National Toxicology Program Technical Report, 215:1-116.

Negishi T, Kawasaki K, Takatori A, Ishii Y, Kyuwa S, Kuroda Y, Yoshikawa Y. (2003). Effects of perinatal exposure to bisphenol A on the behavior of offspring in F344 rats. Environ Toxicol Pharmacol, 14: 99-108.

Newbold RR, Jefferson WN, Padilla-Banks E (2007). Long-term adverse effects of neonatal exposure to bisphenol A on the murine female reproductive tract. Reprod Toxicol, 24: 253-258.

Newbold RR, Padilla-Banks E, Jefferson WN (2009). Environmental estrogens and obesity. Mol Cell Endocrinol, 304: 84-89.

Nikaido Y, Yoshizawa K, Danbara N, Tsujita-Kyutoku M, Yuri T, Uehara N, Tsubura A (2004). Effects of maternal xenoestrogen exposure on development of the reproductive tract and mammary gland in female CD-1 mouse offspring. Reprod Toxicol, 18: 803-811.

Ning G, Bi Y, Wang T, Xu M, Xu Y, Huang Y, Li M, Li X, Wang W, Chen Y, Wu Y, Hou J, Song A, Liu Y, Lai S (2011). Relationship of urinary bisphenol A concentration to risk for prevalent type 2 diabetes in Chinese adults: a cross-sectional analysis. Ann Intern Med, 155: 368-374.

Obata T, Kubota S (2000). Formation of hydroxy radicals by environmental estrogen-like chemicals in rat striatum. Neurosci Lett, 296: 41-44.

Ogiue-Ikeda M, Tanabe N, Mukai H, Hojo Y, Murakami G, Tsurugizawa T, Takata N, Kimoto T, Kawato S (2008). Rapid modulation of synaptic plasticity by estrogens as well as endocrine disrupters in hippocampal neurons. Brain Res Reviews, 57: 363-375.

Olsén L, Lind L, Lind PM (2012). Associations between circulating levels of bisphenol A and phthalate metabolites and coronary risk in the elderly. Ecotoxicol Environ Saf, 80: 179-183.

Pant J, Ranjan P, Deshpande SB (2011). Bisphenol A decreases atrial contractility involving NO-dependent G-cyclase signaling pathway. J Appl Toxicol, 31: 698-702.

Paris F, Balaguer P, Térouanne B, Servant N, Lacoste C, Cravedi JP, Nicolas JC, Sultan C. (2002). Phenylphenols, biphenols, bisphenol-A and 4-tert-octylphenol exhibit alpha and beta estrogen activities and antiandrogen activity in reporter cell lines. Mol Cell Endocrinol, 193: 43-49.

Pastorea A, Federicia G, Bertini E, Piemonte F (2003). Analysis of glutathione: implication in redox and detoxification Clinica Chimica Acta, 333: 19-39.

Patisaul HB, Bateman HL (2008). Neonatal exposure to endocrine active compounds or an ERbeta agonist increases adult anxiety and aggression in gonadally intact male rats. Horm Behav, 53: 580–588.

Petros A, Bennett D, Vallance P (1991). Effect of nitric oxide synthase inhibitors on hypotension in patients with septic shock. Lancet, 338: 1557– 1558.

Pigeolet E, Corbisier P, Houbion A, Lambert D, Michiels DC, Raes M, Zachary D, Ramacle J (1990). Glutathione peroxidase, superoxide dismutase and catalase inactivation by peroxides and oxygen derived free radicals. Mech Ageing Dev, 51: 283-290.

Reilly JJ (2005). Descriptive epidemiology and health consequences of childhood obesity. Best Pract Res Clin Endocrinol Metab, 19: 327-341.

Richter CA, Birnabaum LS, Farabollini F, Newbold RR, Rubin BS, Talsness CE, Vaderbergh JC, Walser-Kuntz, DR, Vom saal FS (2007). In vivo effect of bisphenol A in laboratory rodent studies. Reprod Toxicol, 24: 199-224.

Ritter S (2011). Debating BPA's toxicity. Chem Eng News, 89: 5-13.

Ropero AB, Alonso-Magdalena P, Garcia-Garcia E, Ripoll C, Fuentes E, Nadal A (2008). Bisphenol-A disruption of the endocrine pancreas and blood glucose homeostasis. Int J Androl, 31: 194-200.

Rosenblum ER, Gavaler JS, Van Thiel DH (1989). Lipid peroxidation: a mechanism for alcohol- induced testicular injury. Free Radic Biol Med, 7(5): 569- 577.

Roy P, Venkat RG, Naidu MUR, Usha RP (2005). Recent trends in the nitrergic nervous system. Educational Forum, 37: 69-76.

Rubin BS, Murray MK, Damassa DA, King JC, Soto AM (2001). Perinatal exposure to low doses of bisphenol-A affects body weight, patterns of estrous cyclicity and plasma LH levels. Environ Health Perspect, 109: 675-680.

Ryan BC, Vandenbergh JG (2006). Developmental exposure to environmental estrogens alters anxiety and spatial memory in female mice. Horm Behav, 50: 85-93.

Ryan KK, Haller AM, Sorrell JE, Woods SC, Jandacek RJ, Seeley RJ (2010). Perinatal exposure to bisphenol-A and the development of metabolic syndrome in CD-1 mice. Endocrinology, 151: 2603-2612.

Sajiki J (2001). Decomposition of bisphenol A by radical oxygen. Environ Int, 27: 315-320.

Sax NI (1985). Toxicology of phenols. In: Sax, N.I. (ed.), Dangerous properties of industrial chemicals. Von Nostrand Reinhold, New York, pp.458-1008.

Scherrer-Crosbie M, Ullrich R, Bloch KD, Nakajima H, Nasseri B, Aretz HT, Lindsey ML, Vançon AC, Huang PL, Lee RT, Zapol WM, Picard MH (2001). Endothelial nitric oxide synthase limits left ventricular remodeling after myocardial infarction in mice. Circulation, 104: 1286-1291.

Shah AM, Vallance P, Harrison D (1999). NO in the cardiovascular system. Cardiovasc Res, 43: 507-508.

Shankar A, Teppala S (2011). Relationship between urinary bisphenol A levels and diabetes mellitus. J Clin Endocrinol Metabol, 96: 3822-3826.

Shankar A, Teppala S, (2012). Urinary bisphenol A and hypertension in a multiethnic sample of US adults. J Environ Public Health, 2012, 481641, 5 pages.

Shankar A, Teppala S, Sabanayagam C (2012). Urinary Bisphenol-A levels and measures of obesity: results from the National Health and Nutrition Examination Survey 2003–2008. ISRN Endocrinology, 2012: 965243.

Sian J, Dexter DT, Lees AJ, Daniel S, Agid Y, Javoy-Agid F, Jenner P, Marsden CD (1994). Alterations in glutathione levels in Parkinson's disease and other neurodegenerative disorders affecting the basal ganglia. Annals Neurol, 36: 348-355.

Silver MK, O'Neill MS, Sowers MR, Park SK (2011). Urinary bisphenol A and type-2 diabetes in U.S. adults: data from NHANES, 2003–2008. PLoS ONE, 6: e26868.

Somm E, Schwitzgebel VM, Toulotte A, Cederroth CR, Combescure C, Nef S, Aubert ML, Hüppi PS (2009). Perinatal exposure to bisphenol A alters early adipogenesis in the rat. Environ Health Perspect, 117: 1549-1555.

Sorg DA, Buckner B (1964). A simple method of obtaining venous blood from small laboratory animals. Proc Soc Exp Bio Med, 115: 1131.

Suarez S, Sueira RA, Garrido G (2000). Genotoxicity of the coating lacqure on food cans, bisphenol and hydrolysis products and diglycidyl ether (BADGE), its hydrolysis products and of chlorohydrins of BADGE. Mutat Res, 470: 221-228.

Sugamura K, Keaney JF (2011). Reactive oxygen species in cardiovascular disease. Free Radic Biol Med, 51: 978-992.

Sun Y, Nakashima MN, Takahashi M, Kuroda N, Nakashima K (2002). Determination of bisphenol A in rat brain by microdialysis and column switching high-performance liquid chromatography with fluorescence detection. Biomedical Chromatography, 16: 319-326.

Takahashi O, Oishi S (2001). Testicular toxicity of dietary 2,2- bis(4-hydroxyl phenyl) propane (bisphenol A) in F344 rats. Arch Toxicol, 75: 42-51.

Takayanagi S, Tokunaga T, Liu X, Okada H, Matsushima A, Shimohigashi Y (2006). Endocrine disruptor bisphenol A strongly binds to human estrogen-related receptor gamma (ERRgamma) with high constitutive activity. Toxicol. Lett, 167: 95-105.

Tanabe N, Yoshino H, Kimoto T (2012). Nanomolar dose of bisphenol A rapidly modulates spinogenesis in adult hippocampal neurons. Mol Cell Endocrinol, 351: 317-325.

Tohei A, Koibuchi H, Tamura K, Honda H, Watanabe G, Taya K, and Kogo H (1999). Bisphenol A inhibits testicular functions and increase LH secretion in male rats. Biol Reprod, 60: 202.

Trasande L, Attina TM, Blustein J (2012). Association between urinary bisphenol A concentration and obesity prevalence in children and adolescents. JAMA 308: 1113-1121.

Tribulová N, Okruhlicová L, Bernátová I, Pechánová O (2000). Chronic disturbances in NO production results in histochemical and subcellular alterations of the rat heart. Physiol Res, 49: 77-88.

Tsakiris S, Angelogianni P, Schulpis KH, Stavridis JC (2000). Protective effect of L-phenylalanine on rat brain acetylcholinesterase inhibition induced by free radicals. Clin Biochem, 33: 103-106.

Tsutsui H, Kinugawa S, Matsushima S (2009). Mitochondrial oxidative stress and dysfunction in myocardial remodeling. Cardiovas Res, 81: 449-456.

Tyl RW, Myers CB, Marr MC, Thomas BF, Keimowitz AR, Brine DR, Veselica MM, Fail PA, Chang TY, Seely JC, Joiner RL, Butala JH, Dimond SS, Cagen SZ, Shiotsuka RN, Stropp GD, Waechter JM

(2002). Three-generation reproductive toxicity study of dietary bisphenol A in CD Sprague-Dawley rats. Toxicol Sci, 68: 121-146.

Vandenberg LN, Chahoud I, Heindel JJ, Padmanabhan V, Paumgartten FJ, Schoenfelder G (2010). Urinary, circulating, and tissue biomonitoring studies indicate widespread exposure to bisphenol A. Environ Health Perspect, 118: 1055-1070.

Vaziri ND, Wang XQ, Oveisi F, Rad B (2000). Induction of oxidative stress by glutathione depletion causes severe hypertension in normal rats. Hypertension, 36: 142-146.

Viberg H, Lee I (2012). A single exposure to bisphenol A alters the levels of important neuroproteins in adult male and female mice. Neurotoxicology, 33: 1390-1395.

vom Saal FS, Myers JP (2008). Bisphenol A and risk of metabolic disorders. JAMA, 300: 1353-1355.

vom Saal FS, Hughes C (2005). An extensive new literature concerning low dose effects of bisphenol A shows the need for a new risk assessment. Environ Health Perspect, 113: 926-933.

Wade GN, Schneider JE (1992). Metabolic fuels and reproduction in female mammals. Neurosci Biobehav Rev, 16: 235-272.

Wang T, Li M, Chen B, Xu M, Xu Y, Huang Y, Lu J, Chen Y, Wang W, Li X, Liu Y, Bi Y, Lai S, Ning G (2012). Urinary bisphenol A (BPA) concentration associates with obesity and insulin resistance. J Clin Endocrinol Metab, 97: E223–E227.

Weber JT (2012). Altered calcium signaling following traumatic brain injury. Front Pharmacol, 3: 60.

Wei J, Lin Y, Li Y, Ying C, Chen J, Song L, Zhou Z, Lv Z, Xia W, Chen X, Xu S. (2011). Perinatal exposure to bisphenol A at reference dose predisposes offspring to metabolic syndrome in adult rats on a high-fat diet. Endocrinology, 152: 3049-3061.

Welshons WV, Nagel SC and Vom Saal FS (2006). Large effects from small exposures. III. Endocrine mechanisms mediating effects of bisphenol A at levels of human exposure. Endocrinology, 147: S56-69.

Welshons WV, Thayer KA, Judy BM, Taylor JA, Curran EM, vom Saal FS (2003). Large effects from small exposures. I. Mechanisms for endocrine-disrupting chemicals with estrogenic activity. Environ Health Perspect, 111: 994-1006.

Wetherill YB, Akingbemi BT, Kanno J, McLachlan JA, Nadal A, Sonnenschein C, Watson CS, Zoeller RT, Belcher SM (2007). In vitro molecular mechanisms of bisphenol A action. Reprod Toxicol, 24: 178-198.

WHO (2011). Toxicological and health aspects of bisphenol A: report of joint FAO/WHO expert meeting and report of stakeholder meeting on bisphenol A, November 1–5, 2010, Ottawa, Canada. World Health Organization and Food and Agriculture Organization of the United Nations, Geneva, Switzerland.

Wyse AT, Stefanello FM, Chiarani F, Delwing D, WannmAChEr CM, Wajner M (2004). Arginine administration decreases cerebral cortex acetylcholinesterase and serum butyrylcholinesterase probably by oxidative stress induction. Neurochem Res, 29: 385-389.

Xu X, Tan L, Himi T, Sadamatsu M, Tsutsumi S, Akaike M, Kato N (2011). Changed preference for sweet taste in adulthood induced by perinatal exposure to bisphenol A: a probable link to overweight and obesity. Neurotoxicol Teratol, 33: 458-463.

Xu XH, Zhang J, Wang YM, Ye YP, Luo QQ (2010). Perinatal exposure to bisphenol-A impairs learning-memory by concomitant down-regulation of N-methyl-D-aspartate receptors of hippocampus in male offspring mice. Horm Behav, 58: 326-333.

Ye XB, Pierik FH, Hauser R, Duty S, Angerer J, Park MM, Burdorf A, Hofman A, Jaddoe VWV, Mackenbach, JP, Steegers EAP, Tiemeier H, Longnecker MP (2008). Urinary metabolite concentrations of organophosphorous pesticides, bisphenol A, and phthalates among pregnant women in Rotterdam, the Netherlands: the Generation R study. Environ Res, 108: 260-267.

Yokomaku D, Numakawa T, Numakawa Y, Suzuki S, Matsumoto T, Adachi N, Nishio C, Taguchi T, Hatanaka H (2003). Estrogen enhances depolarization-induced glutamate release through activation of phosphatidylinositol 3-kinase and mitogen-activated protein kinase in cultured hippocampal neurons. Mol Endocrinol, 17: 831-844.

Yu C, Tai F, Song Z, Wu R, Zhang X, He F (2011). Pubertal exposure to bisphenol A disrupts behavior in adult C57BL/6J mice. Environ Toxicol Pharmacol, 31: 88-99.

Zweier JL, Rayburn BK, Flaherty JT, Weisfeldt ML (1987). Recombinant superoxide dismutase reduces oxygen free radical concentrations in reperfused myocardium. J Clin Invest, 80: 1728-1734.

I want morebooks!

Buy your books fast and straightforward online - at one of the world's fastest growing online book stores! Environmentally sound due to Print-on-Demand technologies.

Buy your books online at
www.get-morebooks.com

Kaufen Sie Ihre Bücher schnell und unkompliziert online – auf einer der am schnellsten wachsenden Buchhandelsplattformen weltweit!
Dank Print-On-Demand umwelt- und ressourcenschonend produziert.

Bücher schneller online kaufen
www.morebooks.de

OmniScriptum Marketing DEU GmbH
Heinrich-Böcking-Str. 6-8
D - 66121 Saarbrücken
Telefax: +49 681 93 81 567-9

info@omniscriptum.com
www.omniscriptum.com

www.ingramcontent.com/pod-product-compliance
Lightning Source LLC
Chambersburg PA
CBHW031543210526
45464CB00003B/1133